Erich Neuwirth

Musical Temperaments

Springer-Verlag Wien GmbH

Univ.-Doz. Dr. Erich Neuwirth
Arbeitsgruppe Computergestützte Didaktik
Institut für Statistik, Operations Research und Computerverfahren
Universität Wien, Österreich

Translated from German by Rita Steblin

© 1997 Springer-Verlag Wien

Originally published by Springer-Verlag/Wien in 1997

Typesetting: Camera-ready by the author
Printing and binding: Adolf Holzhausens Nfg. GesmbH, A-1070 Wien
Graphic design: Ecke Bonk

Printed on acid-free and chlorine-free bleached paper
SPIN: 10634843

Die Deutsche Bibliothek – CIP-Einheitsaufnahme

Musical temperaments / Erich Neuwirth. Transl. from the German
by Rita Steblin. – Wien ; New York : Springer
  Dt. Ausg. u.d.T.: Musikalische Stimmungen
  ISBN 3-211-83040-5
  Buch. 1997
  brosch.
CD-ROM. 1997

Additional material to this book can be downloaded from http://extras.springer.com.

ISBN 978-3-211-83040-6   ISBN 978-3-7091-6540-9 (eBook)
DOI 10.1007/978-3-7091-6540-9

Preface

What you are now reading is the written version of an electronic document that explains the mathematical principles for different musical temperaments. The electronic version contains many music examples that you can listen to while you are working with this document at a computer. The written version obviously cannot offer this possibility. It serves therefore merely as a parallel study aid and guide and cannot replace actually working with the electronic text.

Contents

Introduction and Fundamental Properties

Pitch and Frequency

Preliminary Remarks

It is well known that tones consist of periodically recurring phenomena, that is, beats repeating in a regular pattern.The number of repetitions of beats per second is measured in Hertz: 440 Hertz mean 440 beats per second. This number is also called the frequency of a beat. It is also known that a higher frequency produces a higher pitch. A tone of 550 Hertz is higher than a tone of 440 Hertz. The abbreviation for Hertz is Hz.

 At this place in the electronic version of the document you will find several music examples which you can also listen to.

Let us now compare two tones, one of which doubles exactly the frequency of the other. We will choose 264 Hz and 528 Hz.

We notice that both of these tones sound "equal." We also have the feeling that we are hearing the same tone in different "positions." This will be demonstrated more closely by a further example. Consider the following three rows of numbers:

264	528	1056	2112
330	660	1320	2640
396	792	1584	3168

Each number within these three rows is the double of the preceding number. However, the numbers in the second row are not produced by doubling the numbers of the first row. Listening to these tones gives us the feeling that each of these three rows reproduces one tone only, at different heights, but that each row respectively produces a different tone.

If we proceed from a tone with a specific frequency, then the value of this frequency doubled will correspond with – musically speaking – the same tone, only an octave higher.

We will learn more about the basic properties of this phenomenon in the section Frequencies and Intervals.

Frequencies and Intervals

Now we are going to experiment a little with series of tones. For this purpose let us construct the following row of tones:

The first tone has a frequency of 264 Hz (we will explain later why we are beginning with this frequency). The second tone has this frequency multiplied by two, that is 528 Hz. The third tone has this frequency multiplied by three, that is 792 Hz, etc. We will take all multiples of 264 Hz up to the eightfold frequency. The series of tones, which we will hear immediately, has the following frequencies:

1*264, 2*264, 3*264, 4*264, 5*264, 6*264, 7*264, 8*264

264,   528,   792,   1056,   1320,   1584,   1848,   2112

 At this place in the electronic version of the document you will find several music examples which you can also listen to.

By listening to this succession of tones we notice that intervals are formed that are common to classical western music. For example, the first and second tones are separated by the distance of an octave. The second and third tones form a fifth and the third and fourth tones form a fourth. The fourth and fifth tones form a major third and the fifth and sixth tones form a minor third. The intervals from the sixth to the seventh and from the seventh to the eighth tones do not sound particularly "beautiful" and are not commonly found in classical western music.

We have chosen Middle C as the basic keynote of this series (this tone corresponds with 264 Hz). If we mark the other tones – those having higher frequencies – on a keyboard, then we obtain the following picture:

The black dots correspond with musically significant intervals or tones. The lilac-colored dot (for the tone with the sevenfold frequency) represents a tone that does not appear in conventional scales. The position of this dot shows us approximately where this particular tone lies, that is, between which of the "more common" tones.

A study of intervals and frequencies makes the following apparent:

An interval between two tones is dependent only on the relationship between the two frequencies, not on the absolute value of these

 At this place in the electronic version of the document you will find several music examples which you can also listen to.

Since we now know that intervals are dependent only on the relationship between the frequencies of two tones, we can enter the intervals and the corresponding frequency ratios in a table. The number indicated in each case gives the quotients of the higher to the lower frequency.

Interval	Frequency ratio
Octave	2
Fifth	3/2
Fourth	4/3
Major Third	5/4
Minor Third	6/5

This table can be used as in the following example:

If you wish to find the tone that lies exactly a fourth above the tone of 240 Hz, you must calculate the frequency of this new tone as 240 * 4/3 = 320. Thus, the desired tone has a frequency of 320 Hz.

In the previous music examples we heard tones that were played one by one. However, it is also musically interesting to hear two or more tones played simultaneously.

 At this place in the electronic version of the document you will find several music examples which you can also listen to.

In the introduction we read and heard how the rules for calculating musical intervals work. In the following sections we would like to examine how musically significant scales can be built from these intervals. We recommend that the section Pure Tuning be studied first.

Tuning Systems and Frequencies

Musical Scales in Different Tunings

Pure Tuning

Having learned about frequency ratios in the section Frequencies and Intervals, we would now like to build a major scale.

It would be helpful if we could see the octave on a keyboard.

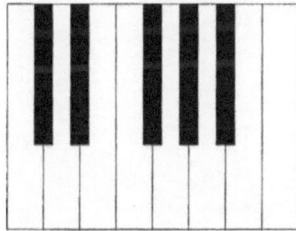

We could begin the scale basically with any arbitrary tone. However, we will begin with Middle C (with a frequency of 264 Hz), because we can then represent the major scale on a piano keyboard in particularly simple terms, namely, on the white keys only.

All further tones are subsequently determined by intervals that they form with the keynote. We have already seen that the octave doubles the frequency of the keynote. This can be represented as follows:

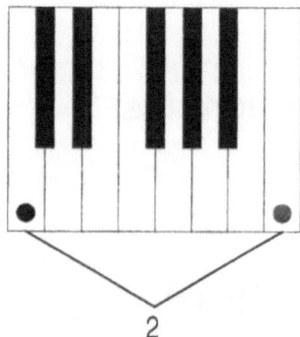

Thus we have defined the frequency of two tones.

The whole scale consists of 8 tones. Thus we still must determine the frequency of the other 6 tones. We learned in the section on Pitch and Frequency that the fifth has a frequency ratio of 3/2. Therefore we can add a further tone to our scale:

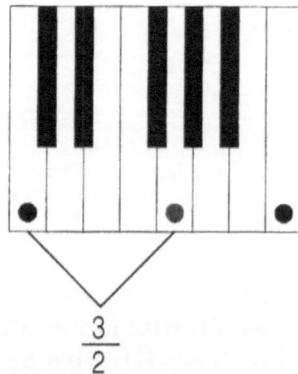

Knowing that the third has a ratio of 5/4 and the fourth a ratio of 4/3, we can add the following two tones:

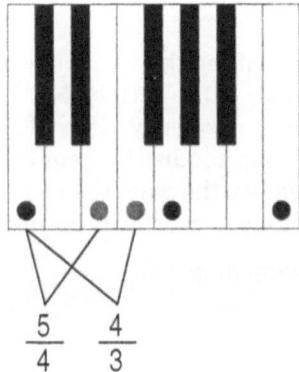

Thus we have now derived the following tones from the keynote:

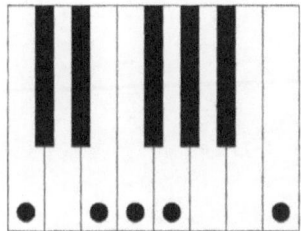

Three tones are still missing. Let us turn our attention first to the two missing high tones. The highest of these two tones cannot be immediately determined from the keynote. But we see that this tone lies exactly a major third above the fifth above the keynote. This becomes clear when we look at the keyboard diagram and observe that one white key and two black keys lie between the fifth and the unknown tone. Two black keys and one white key also lie between the fundamental tone and the third (both intervals are marked in the following diagram):

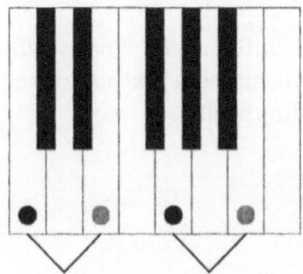

Because of this we can determine the frequency of the new tone by multiplying the frequency of the fifth by 5/4. The frequency of the fifth is 3/2 of the basic frequency. Since 3/2 * 5/4 = 15/8, the frequency of our new tone equals the frequency of the keynote multiplied by 15/8. By the way, the interval between the keynote and the new tone is called the seventh.

Thus we obtain the following diagram:

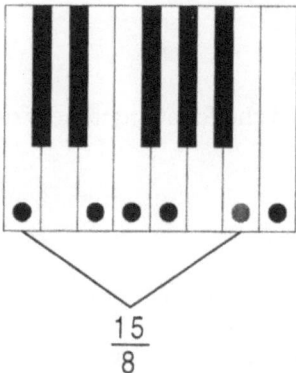

In the same manner, the tone directly under the seventh lies exactly a third above the fourth:

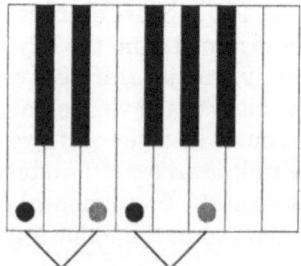

Thus its frequency is calculated as 4/3 * 5/4 = 5/3. The interval with this frequency ratio is called a sixth and we now have arrived at the following diagram:

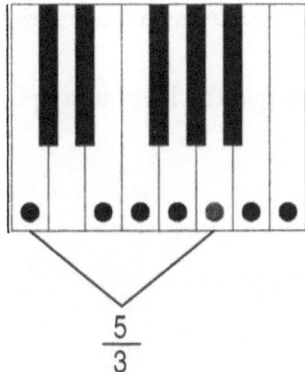

$$\frac{5}{3}$$

We are now missing just a single tone – the second.

The calculation to deduce the second is a little more complicated than that for the other intervals. The following observation is fundamental:

Let us go two fifths upwards from the keynote:

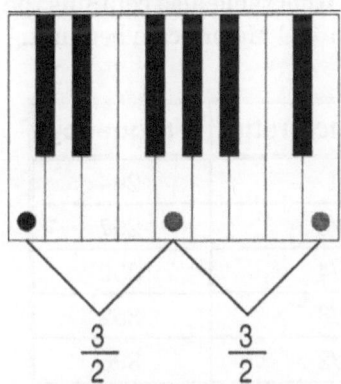

$$\frac{3}{2} \qquad \frac{3}{2}$$

In this way we come to the tone that lies an octave above the second. The frequency of this tone is the basic frequency multiplied by $3/2 * 3/2 = 9/4$.

We observe that this frequency is greater than the basic frequency multiplied by two. And this tone actually lies above the tone with the doubled basic frequency, since this latter tone is the octave. When we go an octave below this new tone (with the basic frequency multiplied by 9/4) we come to the second. Since going up an octave requires a doubling of the frequency, conversely going down an octave requires halving the

frequency. Therefore the second has a frequency ratio of
9/4 * 1/2 = 9/8.

With this we have determined the last missing frequency for
our scale:

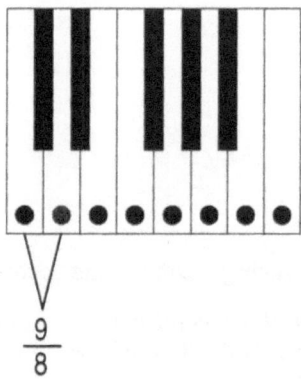

The complete list of frequency ratios for all intervals in pure
tuning looks like this (this table also contains the frequencies
for the tones in the pure C Major scale beginning on Middle C):

Interval	Frequency ratio	Frequency
Keynote	1	264
Second	9/8	297
Third	5/4	330
Fourth	4/3	352
Fifth	3/2	396
Sixth	5/3	440
Seventh	15/8	495
Octave	2	528

 *At this place in the electronic version of the document you will
find several music examples which you can also listen to.*

The basic idea of pure tuning is that all important intervals (the
major and minor thirds and the fifth) are to be expressed by the

simplest possible number ratios. Unfortunately this does not
function as well as one would like it to. The section Intervals
and Triads in Pure Tuning gives more detailed information
about this. The explanation is rather complicated. If you would
rather learn something at this point about another tuning
system, then please turn to the section Pythagorean Tuning.

Intervals and Triads in Pure Tuning

This section is mathematically somewhat more challenging than most of the others. However, if you first skim through this section at least once, you will be able to understand the most important facts in the sections on the other tunings. Above all, this section should help you to understand why tuning systems other than pure tuning are actually necessary.

If you take a major scale in pure tuning and play at the same time the keynote, the major third above and the fifth above, a major triad is formed.

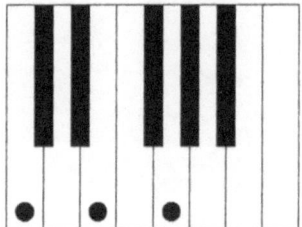

If you click the following diagram, you will be able to hear this chord.

We can form further triads from the tones of the pure major scale if we build a chord consisting of the tone "two scale steps higher" and the tone "four scale steps higher" on each degree of the scale. "One scale step up" here means going from one white key to the next white key.

 At this place in the electronic version of the document you will find several music examples which you can also listen to.

Let us now begin with a more detailed analysis of triads.

Let us see once more the triad formed on the keynote of a pure major scale.

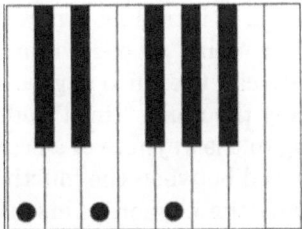

This chord sounds very "pure." The second tone has the 5/4-fold frequency of the keynote and the third tone has the 3/2-fold frequency of the keynote . (see the table of ratios in pure tuning in the appendix "Pictorial explanations").

You also get this chord when you take a basic frequency of 66 Hz and form the fourfold, the fivefold and the sixfold frequency of this basic frequency. The three tones of our chord are therefore the third, fourth and fifth overtones of our basic tone of 66 Hz. If we use only tones in pure tuning, then two further chords will produce a pure major triad.

Let us begin with the tone on the fourth degree and add to this the tones on the sixth and the octave.

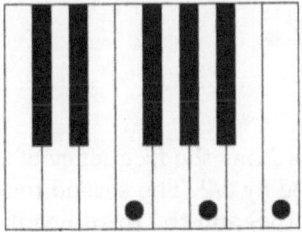

The frequencies of these three tones are the same as the frequency of the keynote of our major scale multiplied by 4/3, 5/3 and 2. Therefore the second tone of the major triad has the frequency of the first tone multiplied by (5/3) / (4/3) = 5/4. The third tone has the frequency of the keynote multiplied by 2 / (4/3) = 3/2. Thus the frequency ratios for the "inner intervals" of this chord are the same as those for a triad on the keynote of a pure major scale (see the table of ratios in pure tuning in the appendix "Pictorial Explanations.")

At this point it will be useful to introduce a new term. So far we have been using names like "third" in order to indicate the degrees of a scale. Thus we call the third degree of the major scale a third. But normally the name "third" indicates this interval without referring to the keynote of a scale. In our triad example, the interval formed between the fourth above the keynote and the sixth above the keynote is also a third. We would now like to name this kind of interval the "inner third" of a triad. Therefore, names like second, third, fourth, etc. normally indicate the degrees of the scale or intervals connected to the keynote of the scale. Names like "inner second," "inner third," "inner fourth," etc., on the other hand, relate to intervals within a chord, whose keynote is not the fundamental tone of the underlying scale.

A major triad is also created when we begin with the fifth degree above the keynote and add to it the seventh and the ninth.

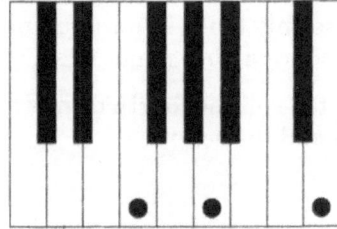

The first tone in this chord has the frequency of the keynote of the major scale multiplied by 3/2. The second tone has this same frequency multiplied by 15/8 and the third has it multiplied by 9/4 (see the table of ratios in pure tuning in the appendix "Pictorial explanations.")

Therefore the second tone has the frequency of the first tone multiplied by (15/8) / (3/2) = 5/4 and the third tone has the frequency of the first tone multiplied by (9/4) / (3/2) = (3/2). Thus the "inner" frequency ratios here are again those of a triad on the fundamental tone of a pure major scale, that is of a pure major triad.

Let us now create a triad on the third below the keynote. This chord consists of the sixth, which has been transposed down an octave, the keynote and the third. This triad is minor.

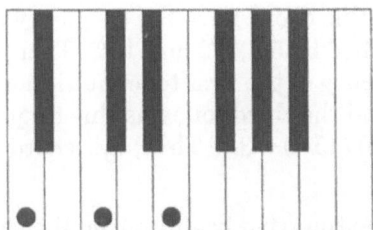

The frequency of the first tone of this triad is the frequency of the keynote of the major scale multiplied by $(5/3) / 2 = 5/6$. The frequency of the keynote is the 5/4-fold of the keynote of the major scale. Therefore the frequency of the second tone is the frequency of the first tone multiplied by $1 / (5/6) = 6/5$ and the frequency of the third tone is the frequency of the first tone multiplied by $(5/4) / (5/6) = 3/2$. These "inner" frequency ratios are characteristic for the minor triad.

For a minor triad the frequency of the second (middle) tone is the frequency of the keynote multiplied by 6/5 and the frequency of the (third) highest tone is the frequency of the keynote multiplied by 3/2. One can also check to see that the three tones of a minor triad have the 10-fold, 12-fold and 15-fold frequency of a common keynote. When, for example, our minor triad consists of tones with the frequencies 220 Hz, 264 Hz and 330 Hz, then these three frequencies are the basic frequency of 22 Hz multiplied by 10, 12 and 15.

Our pure scale also contains another minor chord. When we form a triad based on the third scale degree, then we have a chord made out of the third, fifth and seventh of the pure major scale.

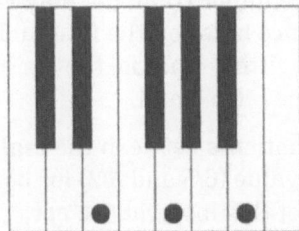

The three tones of this chord have the frequency of the keynote of the scale multiplied by 5/4, 3/2 and 15/8. Therefore the second tone has the frequency of the first tone multiplied by (3/2) / (5/4) = 6/5 and the third tone has this frequency multiplied by (15/8) / (5/4) = 3/2. Thus, this chord is really a minor triad.

Next let us examine the triad beginning on the second degree.

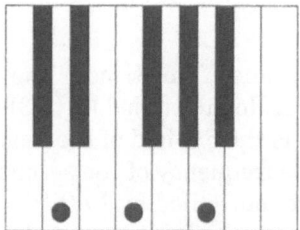

This chord does not sound very pure. Why is this?

The three tones of this triad have the frequency of the keynote of the major scale multiplied by 9/8, 4/3 and 5/3 (see the table of ratios in pure tuning in the appendix "Pictorial explanations.") Therefore the frequency of the second and third tones of this chord amount to the frequency of the first tone of this chord multiplied by (4/3) / (9/8) = 32 / 27 and (5/3) / (9/8) = 40/27. If this chord were a pure minor triad, then the frequency ratios would have to read 6/5 and 3/2. Let us calculate these values as decimal numbers:

32/27 = 1.1852   6/5 = 1.2000

40/27 = 1.4815   3/2 = 1.5000

In comparison with a pure minor triad, not only is the "inner third" – that is, the distance between the first and the second tone – but also the "inner fifth" – that is, the distance between the first and the third tone – too small.

When we calculate the quotients between the real value (32/27 and 40/27) and the ideal value (6/5 and 3/2) for both the inner third and the inner fifth of this interval, we get

(6/5) / (32/27) = 81/80 = 1.0125

(3/2) / (40/27) = 81/80 = 1.0125

This factor 81/80 = 1.0125 is called the syntonic comma (not to be confused with the Pythagorean comma, about which we will speak later).

When, as an experiment, we reduce the value of the second in pure tuning by this syntonic comma, then we obtain a second with a frequency ratio of (9/8) / (81/80) = 10/9 instead of a second with a frequency ratio of 9/8.

 At this place in the electronic version of the document you will find several music examples which you can also listen to.

We would now like to examine which of the two seconds – 9/8 or 10/9 –produce the musically more significant result.

Let us consider the following table:

Interval	Frequ. ratio with the keynote	Frequ. ratio with the inner second	Frequ. ratio with the inner third	Frequ. ratio with the inner fifth
Keynote	1	9/8	5/4	3/2
Second	9/8	10/9	32/27	40/27
Third	5/4	16/15	6/5	3/2
Fourth	4/3	9/8	5/4	3/2
Fifth	3/2	10/9	5/4	3/2
Sixth	5/3	9/8	6/5	3/2
Seventh	15/8	16/15	6/5	64/45
Octave	2	9/8	5/4	3/2

What do the numbers in the columns of this table mean?

- Frequency ratio with the keynote
  shows once again the frequency ratios that we derived in the section Pure Tuning.
  The third above the keynote of the scale has, for example, the frequency of the keynote multiplied by 5/4.

- Frequency ratio with the inner second
  shows the ratio between the frequency of a tone and that of
  the tone a step higher in the pure major scale.
  Proceeding from the third with a ratio of 5/4 with the
  keynote, the tone a step higher, that is the fourth, has the
  ratio of 4/3 with the keynote. Thus, these two tones have a
  ratio with each other of $(4/3) / (5/4) = 16/15$

- Frequency ratio with the inner third
  shows the ratio between the frequency of a tone and that of
  the tone two steps higher in the pure major scale.
  Proceeding from the third with a ratio of 5/4 with the
  keynote, the tone two steps higher, that is the fifth, has the
  ratio of 3/2 with the keynote. Thus, these two tones have a
  ratio with each other of $(3/2) / (5/4) = 6/5$

- Frequency ratio with the inner fifth
  shows the ratio between the frequency of a tone and that of
  the tone four steps higher in the pure major scale.
  Proceeding from the third with a ratio of 5/4 with the
  keynote, the tone four steps higher, that is the seventh, has
  the ratio of 15/8 with the keynote. Thus, these two tones
  have a ratio with each other of $(15/8) / (5/4) = 3/2$

If we now summarize what we have learned about the triads
that can be formed out of the tones of the pure major scale, then
we notice that there are different types of triads.

The triads on the keynote, fourth and fifth have the same inner
intervals, namely a pure major third and a pure fifth (with
reference to the keynote tone of the triad).

The triads on the third and sixth likewise have the same inner
frequency ratios: a pure minor third and a pure fifth (with
reference to the keynote of the triad).

The triad on the seventh consists of a pure minor third and an
interval that we have not yet examined in detail. This interval
is called the diminished fifth; it will not concern us further. By
the way, this type of triad is also called the diminished triad.

A problem is created by the triad on the second. The two inner
intervals are an almost pure minor third and an almost pure
fifth. Both intervals are a syntonic comma too small. For this
reason this triad sounds out of tune. We can try to fix this
impurity by lowering the keynote of this chord by a syntonic

comma, leaving the two higher tones of this chord unchanged. By this means the two "impure" intervals, namely the somewhat flat third and the somewhat flat fifth, are enlarged to the correct intervallic distance. In this manner we obtain a pure minor triad on the second degree of the pure scale.

 At this place in the electronic version of the document you will find several music examples which you can also listen to.

The triad with the modified second sounds purer than the triad with the "original" second. Thus, one could mistakenly believe that the problem with the triad on the second could be solved merely by lowering the second by a syntonic comma. This does not work for the simple reason that this tone (the second) appears in two further triads. Altogether, the second in the major scale appears in three triads:

in the triad on the second

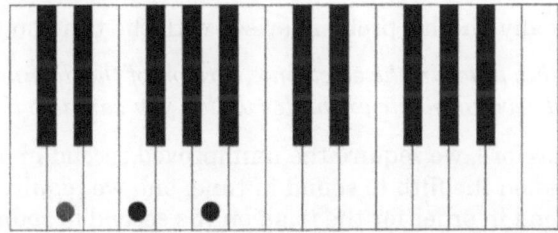

in the triad on the fifth

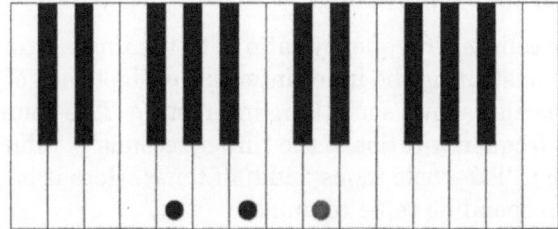

and in the triad on the seventh

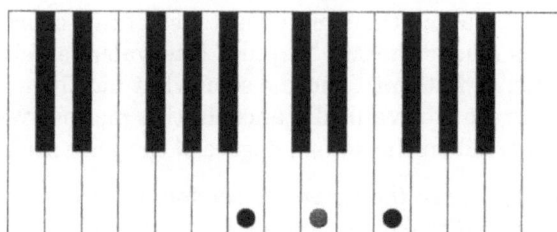

The triad on the second itself is – as we have just seen– "improved" by modifying the second itself. The triad on the seventh presents no special problem, since as a diminished triad it does not sound especially "well tuned" and this effect is not particularly increased by modifying the second.

At this place in the electronic version of the document you will find several music examples which you can also listen to.

A really audible problem arises with the triad on the fifth.

At this place in the electronic version of the document you will find several music examples which you can also listen to.

Therefore, we require the unimproved second in order for the triad on the fifth to sound in tune, and we require the improved second in order for the triad on the second to sound in tune. However, the scale cannot contain both tones at the same time, and thus this creates an insoluble problem in the area of pure tuning. This is one of the reasons why there are tuning systems other than pure tuning.

The column "Frequency ratio with the inner second" in the table demonstrating the inner intervals of the triads of the major scale also shows something interesting. This column contains the frequency ratios of the "inner seconds" – intervals that are also called whole tones and half tones. Here is once again the corresponding table columns:

Interval	Whole tone or half tone	Frequency ratio with the inner second
Keynote	Whole tone	9/8
Second	Whole tone	10/9
Third	Half tone	16/15
Fourth	Whole tone	9/8
Fifth	Whole tone	10/9
Sixth	Whole tone	9/8
Seventh	Half tone	16/15
Octave	Whole tone	9/8

This table shows that pure tuning contains two kinds of whole tones, namely the 9/8-whole tone and the 10/9-whole tone. The difference (or rather the difference ratio) between these two whole tones is something that we already know:
$(9/8) / (10/9) = 81/80$, that is, the syntonic comma.

We also see that the pure major third is made up of a larger and a smaller whole tone. The minor third is made up of a larger whole tone and a half tone.

We see moreover that the major triad contains a major third between its first and second tones and a minor third between its second and third tones. The minor triad also contains these two intervals, only in the reversed order. For the minor triad the distance between the first and second tones is a minor third and the distance between the second and third tones is a major third. It could also be said that the minor triad is a major triad "upside down."

We have tried to solve the problem of the triad on the second by lowering the second by a syntonic comma. For the above table this would mean in effect that the 10/9-whole tone and the 9/8-whole tone would exchange places on the first and second steps of the scale. Thus we would begin with a 10/9-whole tone followed by a 9/8-whole tone. We have seen that we can solve one problem (triad on the second) with this improved tone, but in doing so we create a new problem (triad on the fifth).

Moreover, it is rather unsatisfactory that pure tuning should contain two different kinds of whole tones. Many problems would be solved much more easily if there were only one kind of whole tone. In the further sections we will see how different tuning systems try to deal with the problem just described.

For further reading we now recommend the section Pythagorean Tuning.

Pythagorean Tuning

If you have worked through the section on "Intervals and Triads in Pure Tuning" then you will already know some of the information that will be presented in this section. In the following, we will first deal with some of the properties of pure tuning in greater detail. You should not skip over this material even if some of it is already known to you.

In the section Pure Tuning we examined the topic of pure tuning in detail. This tuning system has the following problem:

The distance from the keynote to the second and from the second to the third actually involves the same interval:

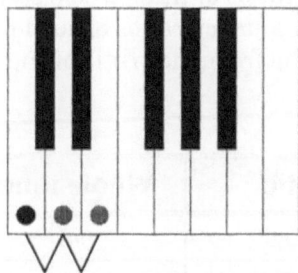

The second corresponds with a frequency ratio of 9/8 and the third with a frequency ratio of 5/4. Therefore the frequency ratio of the second to the third is (5/4) / (9/8) = 40/36 = 10/9. But this interval should be the same interval as the "normal" second, corresponding with the ratio 9/8. It is obvious that not all of the seconds (or, in other words, whole tones) that occur in pure tuning are equal. By whole tone we mean of course a second.

If we examine the piano keyboard, then we will see that several whole tones (or seconds) occur. These appear between the following steps (meaning tones that lie next to each other) of the major scale:

1-2, 2-3, 4-5, 5-6, 6-7

Whole tones are found on a keyboard wherever a black key lies between two white keys.

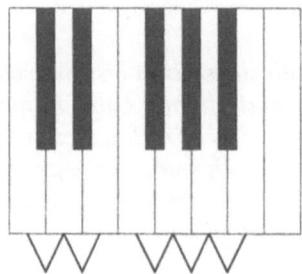

If we take the values for the frequencies of the single steps given in the section Pure Tuning, then we can calculate the values for the frequency ratios of these whole steps. For this we must simply divide the frequency ratios of such neighboring tones (as they relate to the frequency ratio of the keynote) into each other.

Step	Ratio	Whole tone
1-2	9/8 / 1	9/8
2-3	5/4 / 9/8	10/9
4-5	3/2 / 4/3	9/8
5-6	5/3 / 3/2	10/9
6-7	15/8 / 5/3	9/8

We see therefore that there are 5 whole tones in the pure scale and that they are not all the same. There are two different types of whole tones: the whole tones between steps 1-2, 4-5 and 6-7 have a frequency ratio of 9/8; the whole tones between steps 2-3 and 5-6 have a frequency ratio of 10/9.

For the sake of completeness, we would still like to calculate the frequency ratios that belong to the half tones between steps 3-4 and 7-8. (We recognize half tones on the keyboard when no other key lies between the two keys involved.)

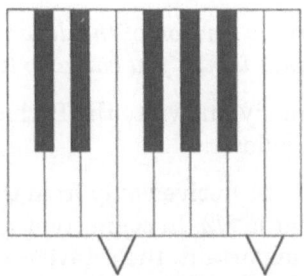

We can see the corresponding frequency ratios in the following table:

Step	Ratio	Half tone
3-4	4/3 / 5/4	16/15
7-8	2 / 15/8	16/15

Thus, these two half tones have the same frequency ratio. Ideally two half tones should produce exactly one whole tone, that is, the whole tone should have the frequency ratio $(16/15) * (16/15) = 256/225$. The decimal notation of this fraction is $256/225 = 1.138$. The two whole tones in pure tuning have a decimal notation of $9/8 = 1.125$ and $10/9 = 1.111$. Thus, in pure tuning not only are there different whole tones, but two half tones are, besides, more than each of these two whole tones.

Let us look at to these three basic tones proceeding from Middle C:

Tone	Ratio	Frequency
Keynote	1	264
Half tone	16/15	281.6
Minor whole tone	10/9	293.3
Major whole tone	9/8	297

At this place in the electronic version of the document you will find several music examples which you can also listen to.

These problems were already known to the Pythagoreans. The solution they chose is as follows:

The complete scale was to be derived only from the frequency ratio of the pure fifth, that is 3/2. In connection with this, the following observation is important: the octave is completed by adding a fourth to the fifth. In other words, the fourth is formed when you take the upper octave and go down a fifth. Therefore we can also use fourths in our system and still remain true to the basic principle of deriving everything from the fifth.

In pure tuning, which intervals are derived directly from the fifth? Answer: the fifth, fourth and second. Thus we can take over the following tones from the pure tuning system:

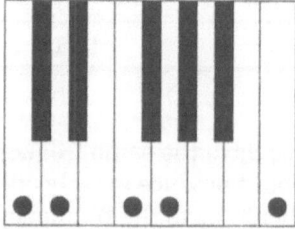

The tones that are still missing are likewise not very difficult to determine:

We obtain the missing third by adding a further second upwards from the second degree. Because the second can be described as "two fifths above and then an octave below," we remain true to our principle of deriving everything from the fifth.

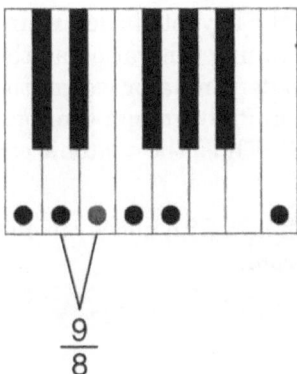

$$\frac{9}{8}$$

Proceeding from the basic frequency, this results in a frequency ratio of 9/8 * 9/8 = 81/64. In pure tuning we had the ratio of 5/4 = 80/64 for this third.

When we compare the third in Pythagoreantuning with the third in pure tuning, and calculate the corresponding frequency ratio, we get (81/64) / (80/64) = 81/80. The musically important question is: Can such a small difference be heard?

At this place in the electronic version of the document you will find several music examples which you can also listen to.

The interval with the frequency ratio 81/80 = 1.025 is called the syntonic comma. The difference between the pure third and the Pythagorean third is equal therefore to a syntonic comma.

The tones that are still missing in our major scale are the sixth and the seventh. We can arrive at the sixth most simply by taking the fifth above the second.

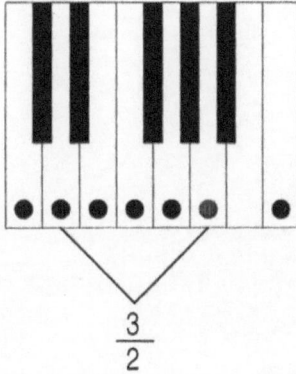

$$\frac{3}{2}$$

The frequency ratio with the keynote is determined as
3/2 * 9/8 = 27/16. In pure tuning this ratio had the value 5/3.
When raised to a common denominator we get the value 81/48
in pure tuning and 80/48 in Pythagorean tuning. The quotient
of these two ratios is 81/80. Thus, the syntonic comma appears
again here too.

We determine the seventh as a fifth above our new third.We see
this in the following diagram:

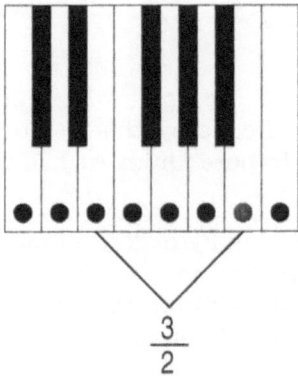

We get the value of 81/64 * 3/2 = 243/128 for this frequency
ratio. In pure tuning the seventh has a value of 15/8 = 240/128.
The quotient of both of these ratios is again 243/240 = 81/80,
thus the syntonic comma again.

If we combine all these values in Pythagorean tuning and
compare them with the corresponding values in pure tuning we
get the following table (the intervals that have different values
in pure and Pythagorean tunings are marked by asterisks):

Interval	Pythagorean	Pure
Keynote	1	1
Second	9/8	9/8
Third*	81/64	5/4
Fourth	4/3	4/3
Fifth	3/2	3/2
Sixth*	27/16	5/3
Seventh*	243/128	15/8
Octave	2	2

If we calculate these fractions in decimal notation we get the following table:

Interval	Pythagorean	Pure
Keynote	1.0000	1.0000
Second	1.1250	1.1250
Third*	1.2656	1.2500
Fourth	1.3333	1.3333
Fifth	1.5000	1.5000
Sixth*	1.6875	1.6667
Seventh*	1.8984	1.8750
Octave	2.0000	2.0000

If, proceeding from this table, we calculate the frequencies of the single tones (in C Major), then we obtain the following values:

Interval	Pythagorean	Pure
Keynote	264.00	264
Second	297.00	297
Third*	334.13	330
Fourth	352.00 .	352
Fifth	396.00	396
Sixth*	445.50	440
Seventh*	501.19	495
Octave	528.00	528

 At this place in the electronic version of the document you will find several music examples which you can also listen to.

We notice clearly audible differences between these two tunings.

In the section Intervals and Triads in Pythagorean Tuning you will learn more about the nature of Pythagorean tuning.

If you would rather learn more about other tuning systems at this point, then go to the section Meantone Tuning.

Intervals and Triads in Pythagorean Tuning

*This section is mathematically somewhat more challenging than
some of the others. However, if you first skim through this
section at least once, you will be able to understand the most
important facts in the sections about the other tunings. Above
all, this section should help you to understand why pure tuning
was changed to Pythagorean tuning.*

If you take a major scale in Pythagorean tuning and play at the
same time the keynote, the major third above and the fifth
above, a major triad is formed.

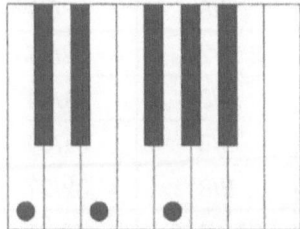

This chord sounds more out of tune than the same chord in pure
tuning.

In a similar way to how we built triads out of the pure major
scale, we can form yet more triads from the tones of the
Pythagorean major scale: we must simply build a chord on each
tone of the scale consisting of that tone, the tone "two scale
steps higher" and the tone "four scale steps higher". "One scale
step up" here again means going from one white key to the next
white key.

*At this place in the electronic version of the document you will
find several music examples which you can also listen to.*

Let us begin now with a more detailed analysis of all these
triads.

To this end, let us consider the following table of inner interval
ratios.

Frequency Ratios, Pythagorean Tuning (as Fractions)

Interval	Frequ. ratio with the keynote	Frequ. ratio with the inner second	Frequ. ratio with the inner third	Frequ. ratio with the inner fifth
Keynote	1	9/8	81/64	3/2
Second	9/8	9/8	32/27	3/2
Third	81/64	256/243	32/27	3/2
Fourth	4/3	9/8	81/64	3/2
Fifth	3/2	9/8	81/64	3/2
Sixth	27/16	9/8	32/27	3/2
Seventh	243/128	256/243	32/27	1024/729
Octave	2	9/8	81/64	3/2

The next table shows the same values as decimal numbers, not as fractions:

Frequency Ratios, Pythagorean Tuning

Interval	Frequ. ratio with the keynote	Frequ. ratio with the inner second	Frequ. ratio with the inner third	Frequ. ratio with the inner fifth
Keynote	1.0000	1.1250	1.2656	1.5000
Second	1.1250	1.1250	1.1852	1.5000
Third	1.2656	1.0535	1.1852	1.5000
Fourth	1.3333	1.1250	1.2656	1.5000
Fifth	1.5000	1.1250	1.2656	1.5000
Sixth	1.6875	1.1250	1.1852	1.5000
Seventh	1.8984	1.0535	1.1852	1.4047
Octave	2.0000	1.1250	1.2656	1.5000

If you wish, you may also compare this table with the corresponding table for pure tuning.

Upon listening to these triads, one notices that the musically important triads on the keynote, fourth and fifth in Pythagorean tuning sound relatively out of tune. Because of this it was necessary to search for other tuning systems. One of these alternative tuning methods is meantone tuning.

Meantone Tuning

A very difficult problem in pure tuning runs as follows:

In the ideal case, after 4 fifths are built up and the top fifth is then dropped down two octaves, a third should result.

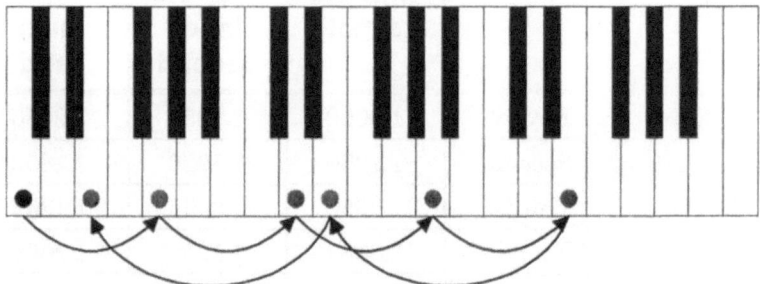

We have already encountered this problem in the construction of Pythagorean tuning. We noticed there that the pure third is not produced when two rising seconds are added together. Since a second equals 2 upwards-rising fifths taken down an octave, 2 seconds should equal 4 rising fifths taken down 2 octaves. We were able to solve the problem at that point by constructing a Pythagorean third consisting of 2 pure seconds. Two rising seconds (or 4 fifths up followed by 2 octaves down) are somewhat larger than the pure third, namely 81/64 compared with 5/4 = 80/64.

In Pythagorean tuning 4 fifths produce a third because the Pythagorean third that is used is somewhat larger than the pure third.

The difference that arises between a pure third and two pure whole tones is called, as we already know, the syntonic comma. One could also say that pure tuning has solved the problem caused by the third and the whole tones in the following manner: one interval has been enlarged by a syntonic comma, namely the third between the keynote and the third degree.

Another way to exact a third out of 4 upward-rising fifths taken down two octaves is not to enlarge the third, but rather to reduce the fifth. These four successive fifths must end on exactly the same tone as a third and two octaves. A third and two successive octaves produce a frequency ratio of

$5/4 * 2 * 2 = 5$. Therefore, the fifth that we now require must have a frequency ratio that produces a 5 when it is multiplied by itself four times. Thus the new fifth has a frequency ratio that equals the 4th root of 5. This is written mathematically as $5^{1/4}$, which yields a value of 1.4953. The pure fifth has a value of $3/2 = 1.5000$. The new fifth now has the desired trait: 4 of these fifths produce a third after 4 rising fifths have been dropped down 2 octaves. Now we would like to form again a complete major scale. The tuning of this scale is called meantone tuning.

The basic intervals for this new construction are the new fifth and the third. Since the new fifth is created out of the third in any case, we could also say that the complete scale is constructed from the pure third. Let us begin with the third, fifth and octave:

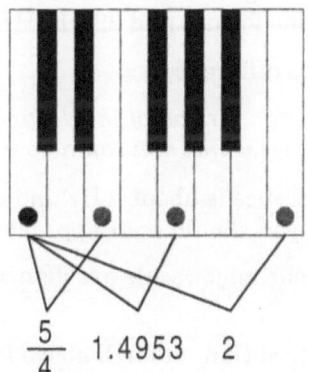

$$\frac{5}{4} \quad 1.4953 \quad 2$$

The next step is to determine the second. We achieve this by taking two of our new fifths in an upwards direction and then by dropping down an octave. Therefore the frequency ratio is $5^{(1/4)} * 5^{(1/4)} / 2 = 1.1180$.

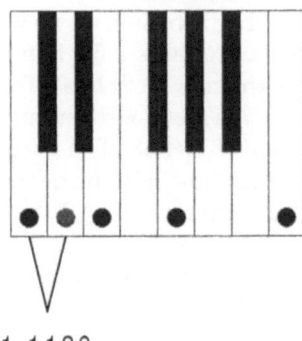

1.1180

As a comparison: the second in pure tuning was 9/8 = 1.1250. The second in Pythagorean tuning was identical with the pure second. When we now calculate the seconds for pure and meantone tuning (proceeding from a basic frequency of 264 Hz), then we get 297 Hz for the former and 295.16 Hz for the latter.

Let us again listen to this difference.

 At this place in the electronic version of the document you will find several music examples which you can also listen to.

Since the frequency difference is about 2 Hz and both tones are played 1.5 seconds long, you should hear approximately 3 beats.

The remaining tones in our major scale are determined as follows:

We have already considered that a fourth above is the same as an octave above taken down a fifth. Thus the fourth must have a frequency ratio of 2 / 1.4953 = 1.3375 in meantone tuning.

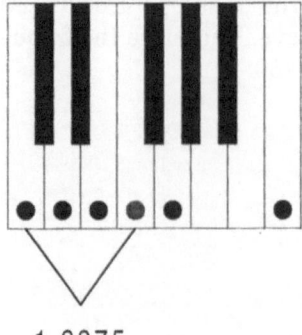

1.3375

This frequency ratio differs from the corresponding ratio for the fourth in both pure and Pythagorean tuning, where it had the value 4/3 = 1.3333.

The two still missing tones are easily obtained with the help of our new fifth: each tone lies a fifth above the second and the third respectively:

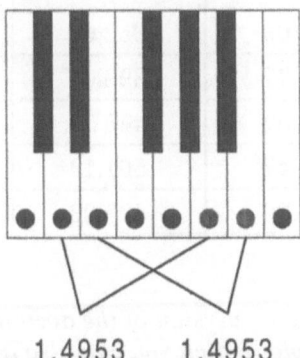

1.4953   1.4953

If we enter the frequency ratios just established in tabular form and compare them with those for pure and Pythagorean tuning, then we obtain the following values:

Interval	Meantone	Pythag.	Pure
Keynote	1.0000	1.0000	1.0000
Second	1.1180	1.1250	1.1250
Third	1.2500	1.2656	1.2500
Fourth	1.3375	1.3333	1.3333
Fifth	1.4953	1.5000	1.5000
Sixth	1.6719	1.6875	1.6667
Seventh	1.8692	1.8984	1.8750
Octave	2.0000	2.0000	2.0000

If, proceeding from this table, we calculate the frequencies of the single tones (in C Major), then we obtain the following values:

Interval	Meantone	Pythag.	Pure
Keynote	264.00	264.00	264
Second	295.16	297.00	297
Third	330.00	334.13	330
Fourth	353.09	352.00	352
Fifth	394.77	396.00	396
Sixth	441.37	445.50	440
Seventh	493.47	501.19	495
Octave	528.00	528.00	528

At this place in the electronic version of the document you will find several music examples which you can also listen to.

There are clearly audible differences between these three tunings.

In the section Intervals and Triads in Meantone Tuning you will learn more about the properties of Pythagorean tuning.

If you would rather learn more about further tuning systems at this point, then proceed to the section Equal Temperament.

Intervals and Triads in Meantone Tuning

This section is mathematically somewhat more challenging than some of the other sections. However, you will be able to understand the most important facts in the sections about other tunings if you first skim through this section at least once. Above all, this section should help you to understand why meantone tuning was a necessary addition to pure tuning and Pythagorean tuning.

A close examination of pure tuning has shown us that there are problems with some triads. The triad on the second in particular sounds very out of tune because the inner fifth of this triad is too small by a syntonic comma and therefore only has a frequency ratio of 40/27 = 1.4815. The corresponding frequency ratio of the pure fifth is 3/2 = 1.5000.

 At this place in the electronic version of the document you will find several music examples which you can also listen to.

This unpleasant characteristic does not occur in Pythagorean tuning. On the other hand, the thirds in the three major triads on the keynote, fourth and fifth are a syntonic comma too high. Thus these triads sound out of tune and not very harmonic.

 At this place in the electronic version of the document you will find several music examples which you can also listen to.

Let us first listen to the major triad on the keynote of a major scale tuned in meantone tuning.

 At this place in the electronic version of the document you will find several music examples which you can also listen to.

This triad sounds very pure, almost as pure as the major triad on the keynote of a major scale in pure tuning.

Just as we formed triads out of pure and Pythagorean major scales, we can also build further triads out of the tones of the meantone major scale: by forming those chords for each tone of the scale that consist of the tone in question, the tone "two scale steps higher" and the tone "four scale steps higher." "One scale step up" here again means going from one white key to the next white key.

 At this place in the electronic version of the document you will find several music examples which you can also listen to.

Let us now begin with a more detailed analysis of all these triads.

In this regard let us consider the following table of inner intervallic ratios.

Frequency Ratios, Meantone Tuning

Interval	Frequ. ratio with the keynote	Frequ. ratio with the inner second	Frequ. ratio with the inner third	Frequ. ratio with the inner fifth
Keynote	1.0000	1.1180	1.2500	1.4953
Second	1.1180	1.1180	1.1963	1.4953
Third	1.2500	1.0700	1.1963	1.4953
Fourth	1.3375	1.1180	1.2500	1.4953
Fifth	1.4952	1.1180	1.2500	1.4953
Sixth	1.6719	1.1180	1.1963	1.4953
Seventh	1.8692	1.0700	1.1963	1.4311
Octave	2.0000	1.1180	1.2500	1.4953

In comparing these triads we notice that the major triads on the keynote, fourth and fifth in meantone tuning are almost exactly as in tune (or euphonic) as those in pure tuning. Moreover, meantone tuning does not have the problem with the out-of-tune minor triad on the second. Furthermore, meantone tuning does not have the main problem that pure tuning had, namely, different whole tones. Therefore, meantone tuning has almost the complete advantage of pure tuning without the severe disadvantage of Pythagorean tuning. Thus it appears that we have found here the "best" tuning system. You can find out why this is not so in the section on Equal Tuning.

Equal Temperament (Tuning)

The circle of fifths is a very important musical concept: if we go up 12 fifths then we should arrive again at the starting point (at a higher range, of course, namely 7 octaves higher). This is illustrated by the following diagram:

The frequency ratio of an octave equals 2. Therefore, the frequency ratio of two tones separated by a distance of 7 octaves has the value $2^7 = 128$.

The frequency ratio of a fifth in pure tuning is $3/2 = 1.5$. Therefore, the frequency ratio of two tones separated by 12 fifths has the value $(3/2)^{12} = 1.5^{12} = 531441/4096 = 129.7463$.

If we start from the same keynote and build up 12 fifths on the one hand and 7 octaves on the other hand, then we obtain two tones whose frequencies are related to each other as follows: $((3/2)^{12}) / 2^7 = 3^{12}/2^{19} = 531441/524288 = 1.013643$. This ratio is called the Pythagorean comma. It is "the amount by which the circle of fifths is not closed in pure tuning." The tone that lies 12 fifths above the starting point is slightly higher than the tone that lies 7 octaves above this same starting point. The pure fifth is therefore a little too large to create a circle of fifths that is completely closed.

At this place in the electronic version of the document you will find several music examples which you can also listen to.

Because Pythagorean tuning uses the same frequency ratio for the fifth as does pure tuning, the problem of the open circle of fifths arises also in the latter system.

What happens to this problem in meantone tuning?

The fifth in meantone tuning has a frequency ratio of $5^{(1/4)}$. Therefore, two tones separated by 12 such fifths have a frequency ratio of $5^3 = 125$. This is a little too low when compared with the tone 7 octaves above the keynote. If we start at the same keynote and go up 12 meantone fifths on the one hand, and then 7 octaves on the other hand, we get two tones with a frequency ratio of $125/128 = 0.9766$. Therefore, the meantone fifth is a little too small for a closed circle of fifths.

At this place in the electronic version of the document you will find several music examples which you can also listen to.

If the circle of fifths is to be closed exactly, how large should the fifth be? Remember, 12 such fifths must equal 7 octaves. We need a number whose 12th power equals 128, thus $128^{(1/12)} = 2^{(7/12)} = 1.4983$. This fifth closes the circle of fifths.

Let us build our scale again, now based on this fifth. If we, as we did earlier, use the fact that the fourth can be created by going "up an octave and then down a fifth," then we can form the first 4 tones of our scale immediately:

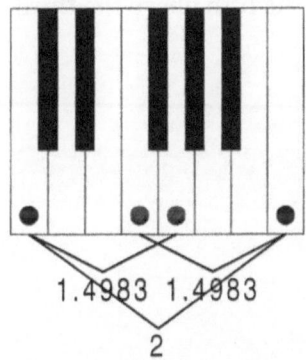

In equal temperament – as in Pythagorean and meantone tuning – the second is formed by the method "two fifths up and then down an octave." Its frequency ratio equals $(2^{(7/12)})^2 / 2 = 2^{(14/12)} / 2 = 2^{(2/12)} = 2^{(1/6)} = 1.1225$. With this value for the frequency ratio of the second, we can form the second of the major scale from the keynote. Using this same value and building up from the second, we can likewise form the third of the major scale.

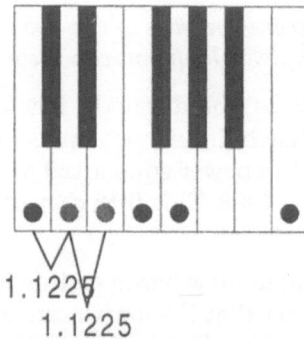

1.1226
1.1225

The two still missing steps of our scale are obtained as the fifth above the second and as the fifth above the third:

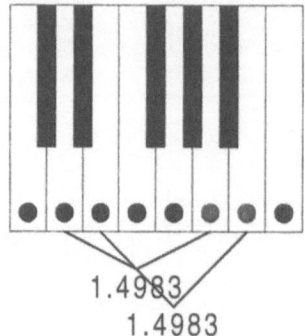

1.4983
1.4983

If we enter these frequency ratios in a table and at the same time compare them with those in meantone, pure and Pythagorean tunings, we obtain the following values:

Interval	Equal	Meantone	Pythag.	Pure
Keynote	1.0000	1.0000	1.0000	1.0000
Second	1.1225	1.1180	1.1250	1.1250
Third	1.2599	1.2500	1.2656	1.2500
Fourth	1.3348	1.3375	1.3333	1.3333
Fifth	1.4983	1.4953	1.5000	1.5000
Sixth	1.6818	1.6719	1.6875	1.6667
Seventh	1.8877	1.8692	1.8984	1.8750
Octave	2.0000	2.0000	2.0000	2.0000

If we start from this table and calculate the frequencies of the single tones (in C Major), we obtain the following values :

Interval	Equal	Meantone	Pythag.	Pure
Keynote	264.00	264.00	264.00	264
Second	296.33	295.16	297.00	297
Third	332.62	330.00	334.13	330
Fourth	352.40	353.09	352.00	352
Fifth	395.55	395.77	396.00	396
Sixth	443.99	441.37	445.50	440
Seventh	498.37	493.47	501.19	495
Octave	528.00	528.00	528.00	528

At this place in the electronic version of the document you will find several music examples which you can also listen to.

There are clearly audible differences between these four tuning systems.

In the section Intervals and Triads in Equal Temperament you will learn more about the nature of equal temperament.

The section labeled Summary gives in condensed form all the important information about the four tuning systems dealt with in this document.

Intervals and Triads in Equal Temperament (Tuning)

As we did for the other tuning systems (pure, Pythagorean and meantone), we would now like to examine for equal temperament the inner intervallic ratios for triads on the various degrees of the major scale.

Let us at first listen to the major triad on the keynote of a major scale tuned in equal temperament.

At this place in the electronic version of the document you will find several music examples which you can also listen to.

This triad sounds rather pure – almost as pure as the major triad on the keynote of a major scale in pure or meantone tuning and, in any case, purer than the triad on the keynote in Pythagorean tuning.

Just as we formed triads out of pure, Pythagorean and meantone major scales, we can also build triads out of the tones of the equally-tempered major scale: by forming those chords for each tone of the scale that consist of the tone in question, the tone "two scale steps higher" and the tone "four scale steps higher." "One scale step up" here again means going from one white key to the next white key.

At this place in the electronic version of the document you will find several music examples which you can also listen to.

Let us now begin with a more detailed analysis of all these triads. In this regard, let us consider the following table of inner intervallic ratios.

Frequency Ratios, Equal Temperament

Interval	Frequ. ratio with the keynote	Frequ. ratio with the inner second	Frequ. ratio with the inner third	Frequ. ratio with the inner fifth
Keynote	1.0000	1.1225	1.2599	1.4983
Second	1.1225	1.1225	1.1892	1.4983
Third	1.2599	1.0595	1.1892	1.4983
Fourth	1.3348	1.1225	1.2599	1.4983
Fifth	1.4983	1.1225	1.2599	1.4983
Sixth	1.6818	1.1225	1.1892	1.4983
Seventh	1.8877	1.0595	1.1892	1.4142
Octave	2.0000	1.1225	1.2599	1.4983

This table shows that the triads on the keynote, fourth and fifth have the same inner frequency ratios. Therefore, all these triads sound the same. Moreover, the problem that some triads sound especially in tune while others sound especially out of tune does not exist. In the same manner, the inner frequency ratios for the triads on the second, third and sixth are all the same. Therefore, all these three triads (all are minor triads) sound exactly the same. Thus, in equal temperament, triads that should theoretically sound the same are in fact the same. This was not the case in any of the other tuning systems (pure, Pythagorean and meantone) that we examined previously.

Furthermore, this table also shows that in equal temperament all of the whole tones in question are of the same size; they have the frequency ratio 1.1225. The half tones in question have the frequency ratio 1.0595. Since $1.0595^2 = 1.1225$, each whole tone in equal temperament has the same size as 2 half tones.

The name "equal temperament" is derived from the fact that the frequency ratios of all similar "inner" intervals on different degrees of the scale, are all the same. In other tuning systems, the frequency ratios for the same inner intervals on different degrees of the scale are often noticably different.

The following paragraph goes beyond the scope of the topic dealt with in this text and gives a brief view of further possible areas of inquiry. Therefore the concepts that appear here will not be explained in as much detail as previous concepts were.

In the course of this text we have not dealt with chromatic tunings and scales. Chromatic scales are formed when half tones are inserted between all the whole tones in a major scale. This makes it possible to transpose; that is, for example, to play a piece of music in D Major instead of E Major. As we have seen, transposition can cause the "sound character" of triads to be noticeably changed. Only equal temperament, on account of its construction, makes it possible that in a transposed piece the character of all triads is retained and only the pitch is altered.

One of the most famous music compositions associated with the problem of tuning is Johann Sebastian Bach's "Well-Tempered Klavier." The first number from this keyboard cycle is played for you at the start of the present document. A tuning system is called "well-tempered" when it is uniform enough so that, for example, one can play a tolerably-sounding D-flat Major (in particular the triad on the keynote of this scale) on an instrument that has been tuned to C Major. This condition is met by equal temperament. "Well-tempered" refers not only to equal temperament but also to every tuning system in which "strange" keys can be played in a musically reasonable manner. It is still not certain even today which tuning system Bach had in mind when he composed the "Well-Tempered Klavier."

In the section Summary you will find a review of all the important material about the four musical tuning systems discussed in this text.

Summary

In this document we are discussing the following 4 important tuning systems:

Pure tuning
Pythagorean tuning
Meantone tuning
Equal temperament

Pure tuning. is the point of departure for all common musical tuning systems. The major scale in pure tuning is formed from the pure third (frequency ratio 5/4) and the pure fifth (frequency ratio 3/2). These two pure intervals (fifth and third) are derived from the overtone series of a tone.

The pure scale has several problems:
Two half tones together do not produce a whole tone..
Two successive seconds do not produce an exact third.
Four fifths up followed by a drop of 2 octaves do not produce an exact third.
12 successive fifths up followed by a drop of 7 octaves do not produce the same starting tone again.

Pythagorean tuning has a higher third than pure tuning, and the Pythagorean fifth is the same as the pure fifth. The Pythagorean third is produced by "adding together" two successive pure seconds. As a consequence of this, two seconds together form a third in Pythagorean tuning. The other problems that we already know from pure tuning (2 half tones not equaling a whole tone, 4 fifths not equaling a third + 2 octaves, 12 fifths not equaling 7 octaves) are not resolved here.

Meantone tuning uses the same third as a pure third, but a fifth that is slightly smaller than the pure fifth. The meantone fifth is just large enough so that 4 such fifths form an interval with the range "two octaves and a pure third." In addition, two seconds together also result in a third in meantone tuning. The other problems that we met with in pure tuning and Pythagorean tuning (2 half tones not equaling a whole tone, 4 fifths not equaling a third + 2 octaves), still remain unsolved.

Equal temperament also uses a fifth that is smaller than the pure fifth. This fifth is however slightly larger than the meantone fifth. The equal temperament third is slightly larger than the pure third (which also equals the meantone third), and

slightly smaller than the Pythagorean third. The equal
temperament fifth is just large enough so that 12 such fifths
produce an interval with the range of 7 octaves, thereby closing
the circle of fifths. In equal temperament two seconds together
produce a third, and two half tones together produce a whole
tone. Also, the last problem that we encountered in the pure
and Pythagorean tuning systems (4 fifths not equaling a third +
2 octaves) is solved in equal temperament. The price paid,
however, is that in equal temperament no interval is equal to an
interval from pure tuning and therfore no interval can be
derived directly from the overtone series or expressed as a
simple fraction.

Appendices

Pictorial explanations

Table of Ratios for Pure Tuning

In the following diagram you see for each tone of the major scale its frequency as the multiple of the frequency of the keynote of this scale.

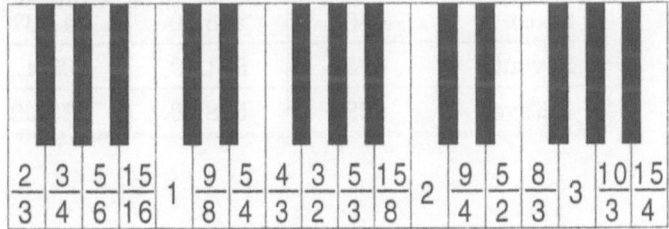

Tables of Frequencies and Intervals

Tones in the Individual Tunings

Degree	Pure	Pythag.	Meantone	Equal
Keynote	264	264.00	264.00	264.00
Second	297	297.00	295.16	296.33
Third	330	334.13	330.00	332.62
Fourth	352	352.00	353.09	352.40
Fifth	396	396.00	395.77	395.55
Sixth	440	445.50	441.37	443.99
Seventh	495	501.19	493.47	498.37
Octave	528	528.00	528.00	528.00

Intervals in the Individual Tunings

Interval	Pure	Pythag.	Meantone	Equal
Keynote	1.0000	1.0000	1.0000	1.0000
Second	1.1250	1.1250	1.1180	1.1225
Third	1.2500	1.2656	1.2500	1.2599
Fourth	1.3333	1.3333	1.3375	1.3348
Fifth	1.5000	1.5000	1.4953	1.4983
Sixth	1.6667	1.6875	1.6719	1.6818
Seventh	1.8750	1.8984	1.8692	1.8877
Octave	2.0000	2.0000	2.0000	2.0000

Frequency Ratios, Pure Tuning (as Fractions)

Interval	Frequ. ratio with the keynote	Frequ. ratio with the inner second	Frequ. ratio with the inner third	Frequ. ratio with the inner fifth
Keynote	1	9/8	5/4	3/2
Second	9/8	10/9	32/27	40/27
Third	5/4	16/15	6/5	3/2
Fourth	4/3	9/8	5/4	3/2
Fifth	3/2	10/9	5/4	3/2
Sixth	5/3	9/8	6/5	3/2
Seventh	15/8	16/15	6/5	64/45
Octave	2	9/8	5/4	3/2

Frequency Ratios, Pythagorean Tuning (as Fractions)

Interval	Frequ. ratio with the keynote	Frequ. ratio with the inner second	Frequ. ratio with the inner third	Frequ. ratio with the inner fifth
Keynote	1	9/8	81/64	3/2
Second	9/8	9/8	32/27	3/2
Third	81/64	256/243	32/27	3/2
Fourth	4/3	9/8	81/64	3/2
Fifth	3/2	9/8	81/64	3/2
Sixth	27/16	9/8	32/27	3/2
Seventh	243/128	256/243	32/27	1024/729
Octave	2	9/8	81/64	3/2

Frequency Ratios, Pure Tuning

Interval	Frequ. ratio with the keynote	Frequ. ratio with the inner second	Frequ. ratio with the inner third	Frequ. ratio with the inner fifth
Keynote	1.0000	1.1250	1.2500	1.5000
Second	1.1250	1.1250	1.1852	1.4815
Third	1.2500	1.0667	1.2000	1.5000
Fourth	1.3333	1.1250	1.2500	1.5000
Fifth	1.5000	1.1111	1.2500	1.5000
Sixth	1.6667	1.1250	1.2000	1.5000
Seventh	1.8750	1.0667	1.2000	1.4222
Octave	2.0000	1.1250	1.2500	1.5000

Frequency Ratios, Pythagorean Tuning

Interval	Frequ. ratio with the keynote	Frequ. ratio with the inner second	Frequ. ratio with the inner third	Frequ. ratio with the inner fifth
Keynote	1.0000	1.1250	1.2656	1.5000
Second	1.1250	1.1250	1.1852	1.5000
Third	1.2656	1.0535	1.1852	1.5000
Fourth	1.3333	1.1250	1.2656	1.5000
Fifth	1.5000	1.1250	1.2656	1.5000
Sixth	1.6875	1.1250	1.1852	1.5000
Seventh	1.8984	1.0535	1.1852	1.4047
Octave	2.0000	1.1250	1.2656	1.5000

Frequency Ratios, Meantone Tuning

Interval	Frequ. ratio with the keynote	Frequ. ratio with the inner second	Frequ. ratio with the inner third	Frequ. ratio with the inner fifth
Keynote	1.0000	1.1180	1.2500	1.4953
Second	1.1180	1.1180	1.1963	1.4953
Third	1.2500	1.0700	1.1963	1.4953
Fourth	1.3375	1.1180	1.2500	1.4953
Fifth	1.4952	1.1180	1.2500	1.4953
Sixth	1.6719	1.1180	1.1963	1.4953
Seventh	1.8692	1.0700	1.1963	1.4311
Octave	2.0000	1.1180	1.2500	1.4953

Frequency Ratios, Equal Temperament

Interval	Frequ. ratio with the keynote	Frequ. ratio with the inner second	Frequ. ratio with the inner third	Frequ. ratio with the inner fifth
Keynote	1.0000	1.1225	1.2599	1.4983
Second	1.1225	1.1225	1.1892	1.4983
Third	1.2599	1.0595	1.1892	1.4983
Fourth	1.3348	1.1225	1.2599	1.4983
Fifth	1.4983	1.1225	1.2599	1.4983
Sixth	1.6818	1.1225	1.1892	1.4983
Seventh	1.8877	1.0595	1.1892	1.4142
Octave	2.0000	1.1225	1.2599	1.4983

Operating Instructions

What are all the things you can do with the mouse?

In this electronic document you will find text passages that are displayed in color. When you click such text passages with the mouse, something usually happens. At some of these passages, like at the one just given, an explanatory text will appear in a secondary window. Sometimes a click of the mouse will also lead you to another section of this document.

Sometimes the electronic document will have numbers that are displayed in color, as for example, 264 Hz. When you click such a marked number, you will hear a tone with the corresponding frequency.

You will also find diagrams at some places in the document that look approximately like the following (to the right is found a loudspeaker symbol):

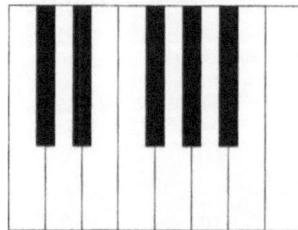

On this kind of diagram you can "play the piano." When you click the single keys you will then hear the corresponding tones.

Sometimes dots appear on several of the keys, as in the following example:

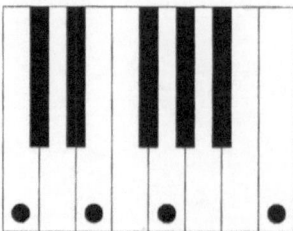

In this case, not all of the keys work: only those that are marked with dots will sound. In some cases only those keys with dots in a particular color will work. The other keys with different colored dots are "silent."

There are several other symbols in this document.

Most of these play intervals or scales when you click the symbols. Here are some examples:

 Plays single tones or intervals

 Plays several tones at the same time (chords)

 Plays a scale

 Plays two scales (in different tunings) staggered with each other

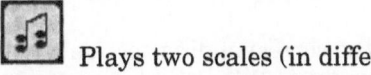 Plays two scales (in different tunings) at the same time through the two stereo channels

Navigating in the Document

This document offers all navigation aids that Windows-Help-Files normally offer. At the upper edge of the main window you see a "Toolbar" with several buttons.

Choosing the first button 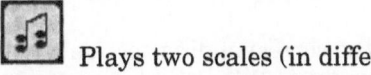 always brings you to the table of contents of the document. The table of contents will appear in a second window, while the first window with the document that you have just been reading remains on the screen. In this manner you can see an overview of the whole contents of the document in one window and read individual portions of the document in the other window.

As in most Windows programs, you can change the size of all the windows. This is accomplished most easily as follows: click the mouse on the "lower right-hand corner" of the window and then, while keeping the mouse depressed, drag it to change the size of the window.

The secondary window with the table of contents at first shows only the main headings of the document. You can see all the other lower level headings if you click the button in the toolbar of the table of contents window. What you will see here is the complete table of contents for the whole document. When you click the button, you will see again only the main headings.

In the table of contents you will see next to the individual chapter headings the following three different symbols:

indicates that the corresponding section has no further subsections, just text. When you click this symbol, the main window will show the corresponding text. Thus, you can jump from section to section in the document by simply clicking the corresponding section heading in the table of contents window with the symbol.

indicates that the corresponding section contains further subsections that are not at this point being displayed in the table of contents window. By clicking this symbol you "open" the corresponding section and the headings of the subsections will then be displayed. At the same time, the symbol changes to .

indicates that the corresponding section contains further subsections. This kind of section does not contain any text; only the further subsections contain the text. The lines after the section heading contain the subsections.

In the toolbar of the table of contents window you will find still further buttons:

brings you from a subsection directly to the start of its main section, that is, from a subordinate chapter to the heading of the corresponding main chapter.

takes you from a section to the immediately following section. In this way you can skip from "section to section."

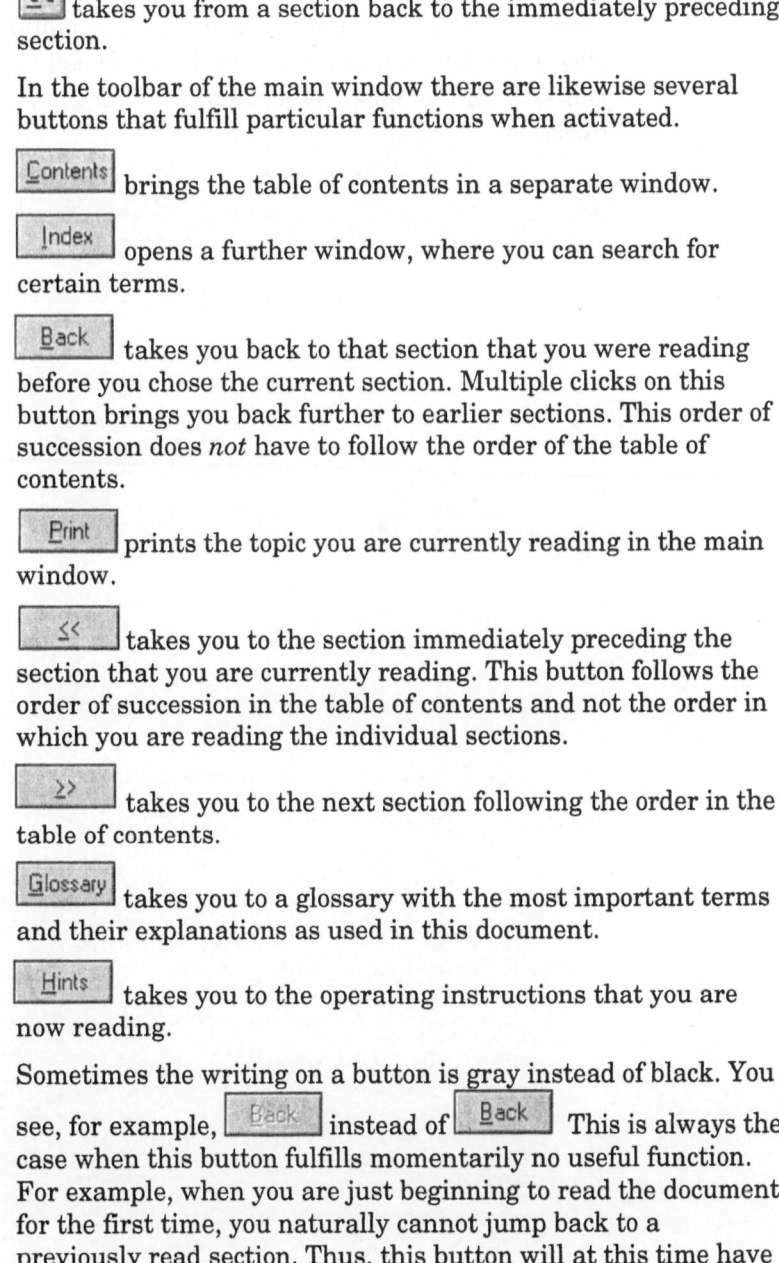

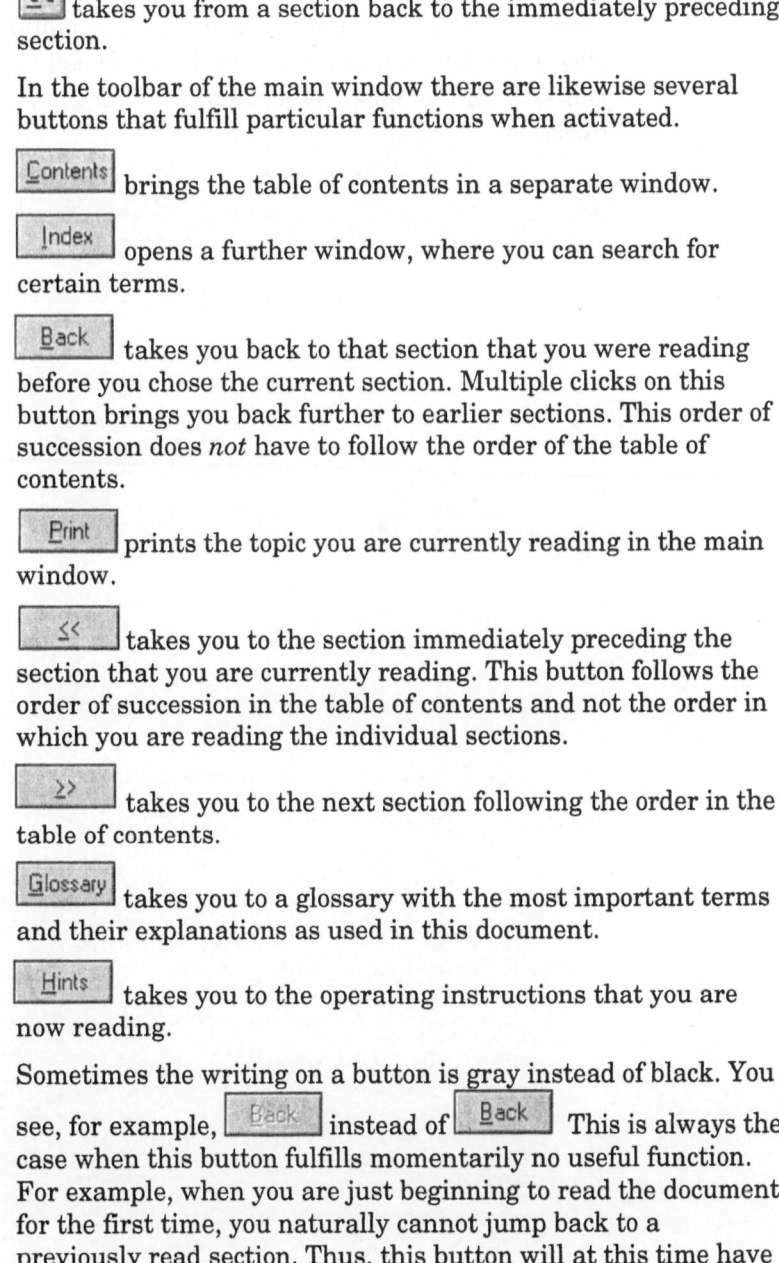 takes you from a section back to the immediately preceding section.

In the toolbar of the main window there are likewise several buttons that fulfill particular functions when activated.

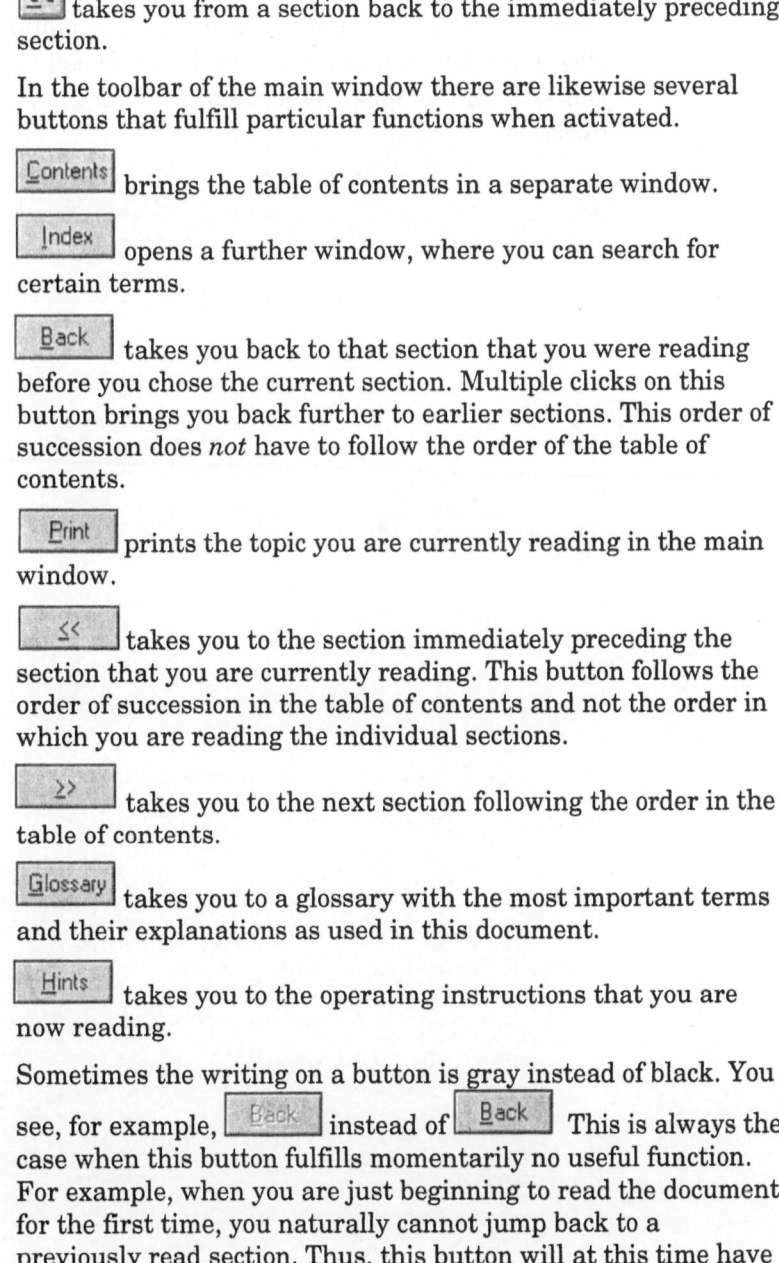 brings the table of contents in a separate window.

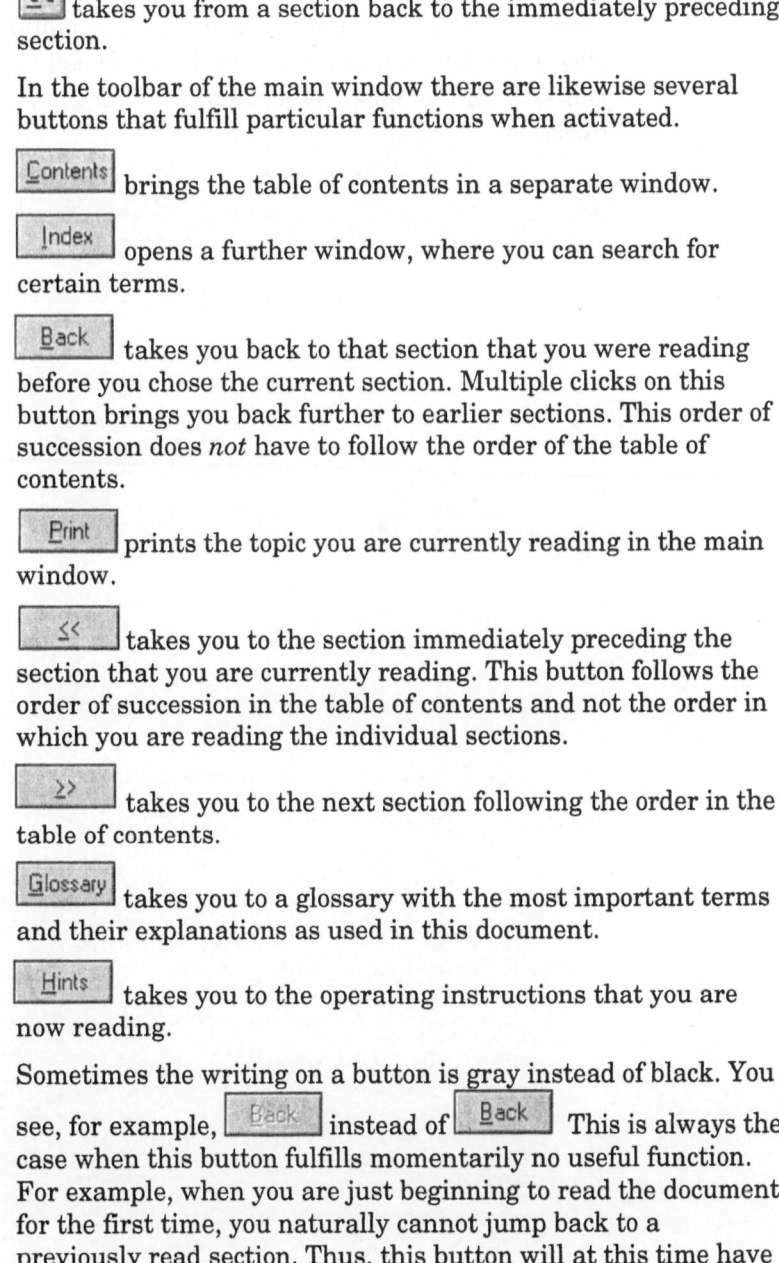 opens a further window, where you can search for certain terms.

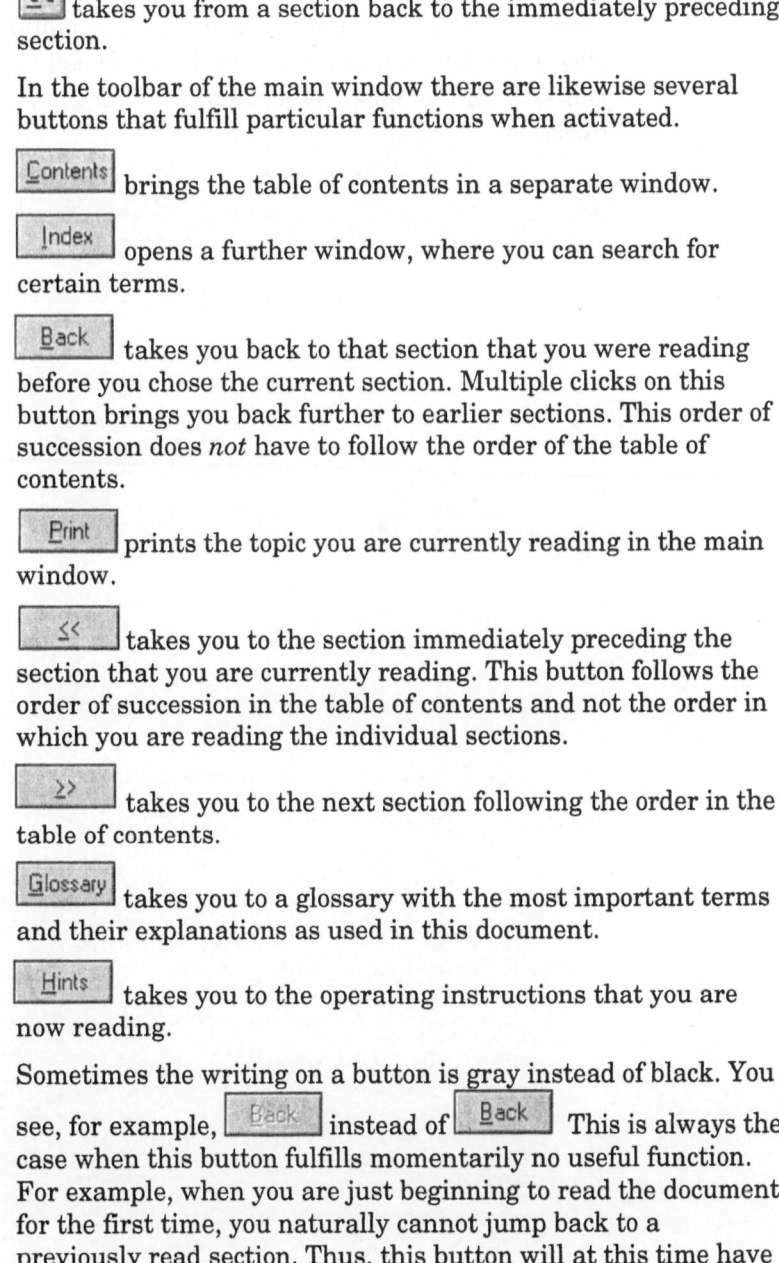 takes you back to that section that you were reading before you chose the current section. Multiple clicks on this button brings you back further to earlier sections. This order of succession does *not* have to follow the order of the table of contents.

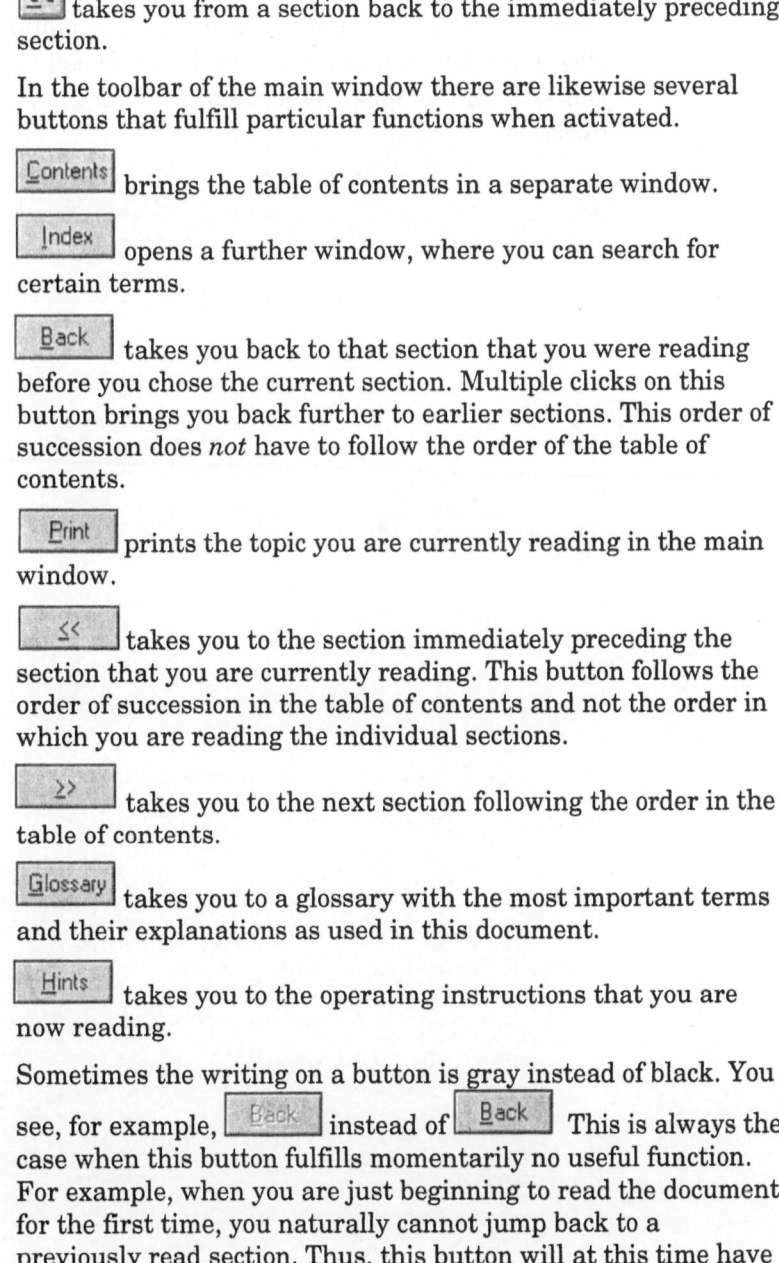 prints the topic you are currently reading in the main window.

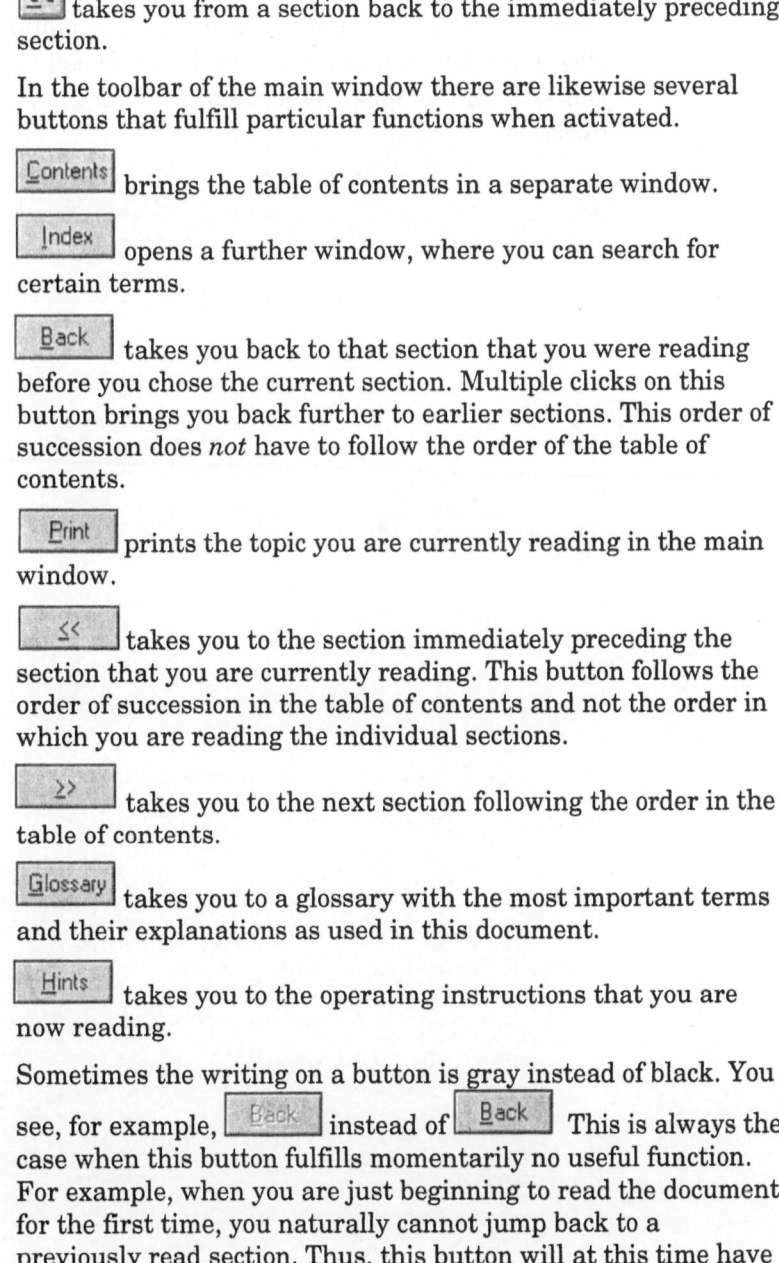 takes you to the section immediately preceding the section that you are currently reading. This button follows the order of succession in the table of contents and not the order in which you are reading the individual sections.

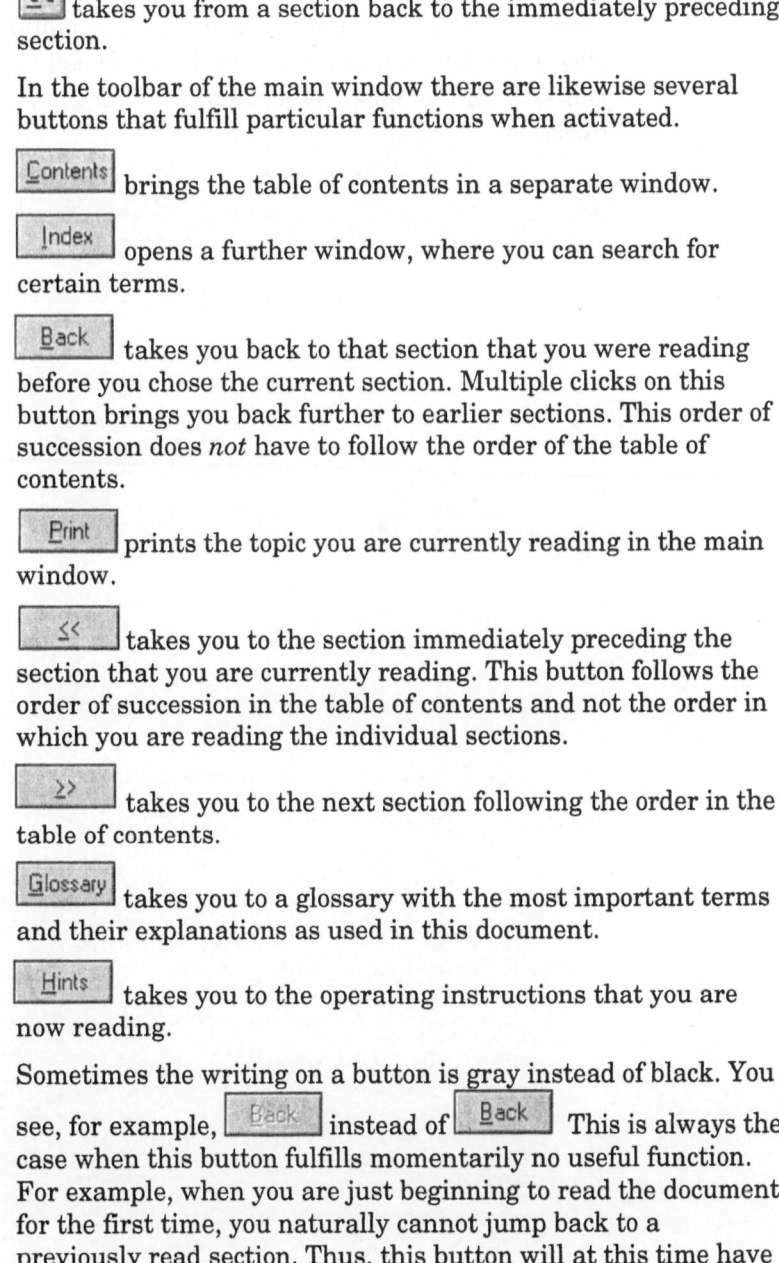 takes you to the next section following the order in the table of contents.

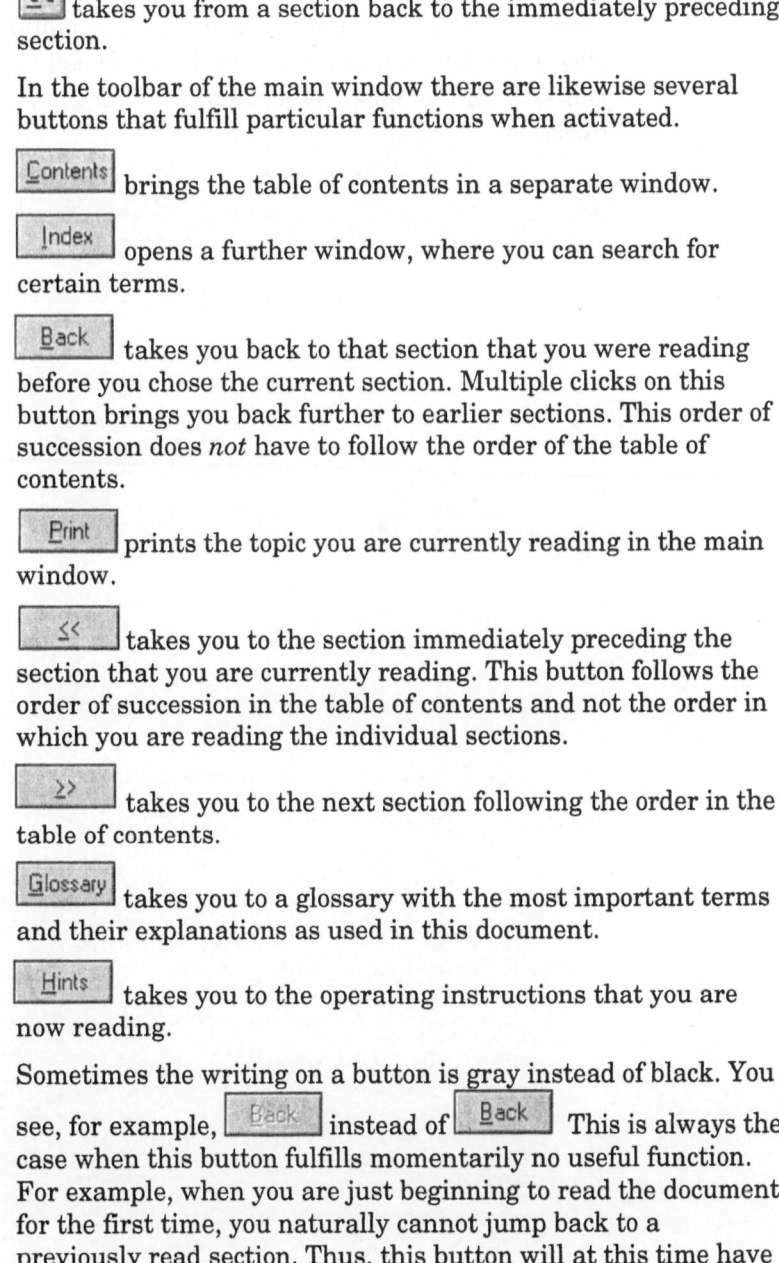 takes you to a glossary with the most important terms and their explanations as used in this document.

 takes you to the operating instructions that you are now reading.

Sometimes the writing on a button is gray instead of black. You see, for example, instead of This is always the case when this button fulfills momentarily no useful function. For example, when you are just beginning to read the document for the first time, you naturally cannot jump back to a previously read section. Thus, this button will at this time have no function to fulfill.

When in the course of studying the document you wish the close one of the opened windows, this is easily accomplished by clicking on the "x mark" – a sign that looks approximately like this – ☒ in the upper right-hand corner of the window in question.

Glossary

Beats

If two tones with nearly the same frequency are played at the same time, then one hears periodic variations in the intensity of sound, known as beats. The frequency of these beats equals the difference between the frequencies of both tones. If a tone of 330 Hz is played at the same time as a tone of 334 Hz, then one should hear beats of 4 Hz: that is, the sound intensity should increase and decrease 4 times per second.

The closer the frequencies, the slower the beats.

Chord

A chord is formed when more than one tone is played at the same time.

Half Tone

One of the fundamental intervals in the building of scales. The same interval as a minor second. A half tone is found on the keyboard when two white keys lying next to each other have no other key (that is, no black key) in between.
Two half tones are marked on the following keyboard:

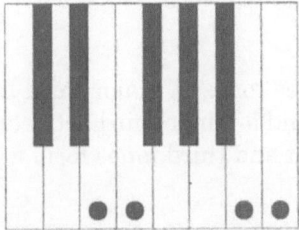

Hertz

The number of beats per second. The abbreviation is Hz.
1 Hz equals one beat per second.

Hz

The abbreviation for Hertz, that is, the number of beats per second.

Inner Interval

In this document interval names like second, third, etc., normally refer to the degrees of the scale, that is, to the intervals of the scale proceeding from the keynote as the starting tone. However, we also study intervals that lie within a chord whose keynote is not the keynote of the underlying scale. We call this kind of interval an inner interval.
Example:
In the triad on the fourth

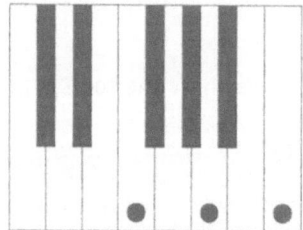

the interval from the first to the second tone of the triad (that is, the interval between the fourth and the sixth notes above the keynote) is a third. We call this third an inner third since it is not the same as the third above the keynote of the scale.

Major Triad

A chord composed of three tones. A major third lies between the first and second tones, and a minor third between the second and third tones. The first and third tones form a pure fifth.

Major Scale

A series of 8 tones. The interval between two successive tones is made up of either a whole tone or a half tone. If we label the whole tones "W" and the half tones "H", then the major scale contains these intervals in the following order: WWHWWWH.

Musical Temperaments

Appendices • 69

Minor Triad

A chord composed of three tones. A minor third lies between the first and second tones, and a major third between the second and third tones. The first and third tones form a pure fifth.

Minor Scale

Series of 8 tones. The interval between two successive tones is made up of either a whole tone or a half tone. If we label the whole tones "W" and the half tones "H", then the minor scale contains these intervals in the following order: WHWWHWW.

Overtone

A tone whose frequency is an integer multiple of the keynote's frequency.
The first overtone has the frequency of the keynote multiplied by two, the second overtone has this frequency multiplied by three, etc.
A tone with a frequency of 660 Hz is, for example, the second overtone of a tone with 220 Hz, as well as the fifth overtone of a keynote with 110 Hz.

Pythagorean Comma

The frequency ratio of $3^{12} / 2^{19} = 531441/524288 = 1.0136$. If you at first go up 12 pure fifths from the keynote and then drop down 7 octaves, the resulting tone will be slightly higher than the starting tone. The exact difference equals the Pythagorean comma.

Not to be confused with the syntonic comma.

Syntonic Comma

Frequency ratio of 81/80 = 1.0125.
Occurs as the difference between the following two tones:
Pure third above a keynote
Two seconds one after the other, starting from the same keynote

Not to be confused with the Pythagorean comma.

Whole Tone

One of the fundamental intervals in the building of scales. The same interval as a major second. A whole tone is found on the keyboard when two white keys lying next to each other are separated by a black key (the black key is not sounded). Two whole tones are marked on the following keyboard:

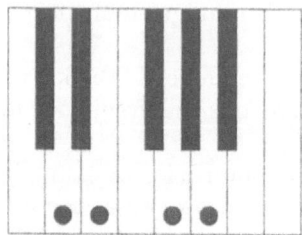

Selected further reading

Easly Blackwood
The Structure of Recognizable Diatonic Tunings
Princeton University Press

Mark Lindley, Ronald Turner-Smith
Mathematical Models of Musical Scales
Verlag für systematische Musikwissenschaft